YOUR KNOWLEDGE HAS VALUE

- We will publish your bachelor's and master's thesis, essays and papers

- Your own eBook and book - sold worldwide in all relevant shops

- Earn money with each sale

Upload your text at www.GRIN.com and publish for free

Bibliographic information published by the German National Library:

The German National Library lists this publication in the National Bibliography;
detailed bibliographic data are available on the Internet at http://dnb.dnb.de .

Imprint:

Copyright © 2009 GRIN Verlag, Open Publishing GmbH
Print and binding: Books on Demand GmbH, Norderstedt Germany
ISBN: 9783640595839

This book at GRIN:

http://www.grin.com/en/e-book/147507/barriers-and-opportunities-in-management-
and-conservation-of-protected

Donal Yeang

Barriers and opportunities in management and conservation of protected areas in Cambodia

GRIN Publishing

Barriers and opportunities in management and conservation of protected areas in Cambodia

Donal Yeang

October 2009

1. Introduction

Protected area has long history in natural reserve in Cambodia and its idea is not new to this nation. It has been introduced for not only management and conservation of ecosystem but also protection of cultural value and landscape. In 1925, 10,800 hectares of forests surrounding Angkor temple was declared as the first national park in Southeast Asia (Wager, 1995; ADB, 2000; ICEM, 2003). To respond to the loss of biodiversity in the nation, conservation and management effort has been made continuously. In 1957, one third of country has been allocated into 173 forest reserves and six wildlife reserves and most of those sites currently declared as the protected areas which offer recreation services to the society along with nature conservation (ADB, 2000). The long delay of civil war caused the management of protected areas to collapse and over the past decade effort was made to restore the protected area systems into practical sense (ICEM, 2003). In 1993, the King Norodom Shihanouk (Father of the present king) issued the decree on a new national protected area system. Ministry of Environment is responsible for the management and development an area of 3,327,200 ha in cooperation with Ministry of Agriculture, Forestry and Fishery (ADB, 2000; ICEM, 2003). According to ADB report (2000), the Royal Decree 126 on "The Creation and Designation of Protection Area" designates 23 protected areas which constitute to 19 percent of the country. In addition, four management categories was defined, namely (1) national parks, (2) wildlife sanctuaries, (3) protected landscapes, and (4) multiple-use management areas. Until February 2008, law on protected areas was approved and it defined the clear roles, obligations and authority of different stakeholders (Cambodian team, 2009). The increase of number of fish sanctuaries and protected forest areas set up through Ministry of Agriculture, Forestry and Fishery bring the national protected area up to 21 percent of the nation (ICEM, 2003).

Even though protected area systems have been put in place, the implementations are still in process and face many challenges. In contrast, there are also some opportunities for protected areas in Cambodia. To have a deep understanding about the current situation of ecosystem conservation in Cambodia, this paper attempts to illustrate some main challenges and opportunities of management and conservation through protected area systems.

2. Biogeographic characteristics of Cambodia

Located in Southeast Asia, Cambodia cover an area of 181 035 square kilometers, with a population of over 14 millions of which about 81 percent lives in the rural areas (Ministry of Rural Development, 2006; Worldbank, 2007). In rural area, large majority of Cambodian earns their income through farming and using natural resources (USAID, 2006). However, 30 percent of its population still lives under the poverty line which means that the people earn less than one US dollar a day (UNDP, 2008). Cambodia is bordered on the north-east by Laos, on the east and south-east by Vietnam, on the south-west by the Gulf of Thailand, and on the west and north-west by Thailand (FAO, 2007). Surrounding by mountains and plateaus (except in the south-east and along the coast), the country has a large alluvial central plain and only a few points exceed 1 000 m in elevation. They are located primarily in the extreme north-east in the Cardamom Mountains and Elephant Mountain (FAO, 2009). Cambodian coastline stretches along the Gulf of Thailand with 64 islands and extensive mangroves and coral reefs (ICEM, 2003). The marine biodiversity is an extraordinary resource of the nation (Worldbank, 2009). The Mekong River and the Tonel Sap, designated as a UNESCO Biosphere Reserve in 1997, likely dominate Cambodian's landscapes (ADB, 2000; ICEM, 2003; UNDP, 2008; Worldbank, 2009). However, the country is believed to have over 130 mammals and more than 500 bird species and about 300 freshwater species (ADB, 2000). Cambodian government estimates that 125 species are endangered. Twenty-eight mammals, 26 birds, 15 reptiles, three amphibians and 32 plants are listed as critically endangered or vulnerable on the IUCN red list (ITTO, 2005). According to the report of ADB (2000) and ICEM (2003), based on the ecological viewpoint Cambodia can be divided into seven unique biodiversity regions namely:

1 South-western Coastal Ranges and Marine Waters: Wet tropical forest including the Cardamom and Elephant Ranges, coastal formations and marine areas generally associated with sandstones. The area has low population densities and is dominated by natural and modified landscapes used for forestry, marine fisheries and the maintenance of biological diversity. Principal ethnic groups living in this area are the Khmer, Pear, Chong and Sóach.

2 Northern Plains: Lowland dry evergreen and associated deciduous forests on sandstones. The region has low population densities and natural and modified landscapes used for forestry, the maintenance of biological diversity, and limited agriculture. Ethnic groups living in this area include Khmer, Pear, Kouy and Stieng.

3 North-eastern Forests: Lowland deciduous forests and limited dry evergreen forest generally associated with sandstones and basalts respectively. The area has low population densities and is dominated by natural and modified landscapes used for forestry, the maintenance of biological diversity, and limited agriculture. Ethnic groups living in this area include the Tampoun, Brao, Rhade, Stieng and Khmer.

4 Kampong Cham: Remnant dry evergreen forests associated with basalts. The region has high population densities and extensive agriculture, plantations and limited forestry. Principal ethnic groups living in this area are the Khmer and Cham.

5 Mekong Delta Region: Characterized by very high population density, these alluvial areas are heavily dominated by agriculture and semi-natural wetlands. Ethnic groups living in this area include Khmer, Cham and some Vietnamese.

6 Tonle Sap Floodplain: This extensive alluvial plain is characterised by unique flooded forest and swamp forests, much of which has been subject to degrading influences. Ethnic groups living in this area are Khmer, Cham and some Vietnamese.

7 North-western region: The Pailin area features lowland evergreen and deciduous forests associated with limestone outcrops. The people living this area are generally Khmer with small numbers of Burmese migrants working in the gem fields. Population densities are higher on the fertile lowland soils of the Battambang Plain, which is highly productive for agriculture.

The definition of region and group of protected areas offer clear view on the resource to be focus for management. Criteria built on biological resources and their use also enables management responses linking protected areas to their surrounding development landscapes (ICEM, 2003).

3. Protected Area System in Cambodia

3.1 Protected Area Definition and Categories by IUCN

Due to ICEM (2003), protected area defines as "an area of land or sea especially dedicated to the protection and maintenance of biological diversity, and of natural and associated cultural resources, and managed through legal or other effective means." Some 120 countries at the Fourth World Congress on National Parks and Protected Areas, held in Caracas, Venezuela in 1992 agreed upon this definition. However, the International Union for Conservation of Nature (IUCN) has revised the definition of the protected area and it was prepared at a meeting on the categories in Almeria, Spain in May 2007. In the new definition, protected area is: "A clearly defined geographical space, recognised, dedicated and managed, through legal or other effective means, to achieve the long-term conservation of nature with associated ecosystem services and cultural values" (Dudley, 2008).The IUCN system classifies all the protected areas into 6 categories based on the principle of management objectives and varying intensities of use (ICEM, 2003).

I Strict Nature Reserve/Wilderness Area: Areas of land and/or sea possessing outstanding or representative ecosystems, geological or physiological features and/or species, available primarily for scientific research and/or environmental monitoring; or large areas of unmodified or slightly modified land, and/or sea, retaining their natural character and influence, without permanent or significant habitation, which are protected and managed so as to preserve their natural condition.

II National Park: Protected Areas Managed Mainly for Ecosystem Conservation and Recreation. Natural areas of land and/or sea, designated to (a) protect the ecological integrity of one or more ecosystems for this and future generations, (b) exclude exploitation or occupation inimical to the purposes of designation of the area, and (c) provide a foundation for spiritual, scientific, educational, recreational and visitor opportunities, all of which must be environmentally and culturally compatible.

III Natural Monument: Protected Areas Managed Mainly for Conservation of Specific Features. Areas containing one or more specific natural or natural/cultural feature which is of outstanding or unique value because of its inherent rarity, representative or aesthetic qualities or cultural significance.

IV Habitat/Species Management Area: Protected Areas Managed Mainly for Conservation Through Management Intervention. Areas of land and/or sea subject to active intervention for management purposes so as to ensure the maintenance of habitats and/or to meet the requirements of specific species.

V Protected Landscape/Seascape: Protected Areas Managed Mainly for Landscape/Seascape Conservation and Recreation. Areas of land, with coast and sea as appropriate, where the interaction of people and nature over time has produced an area of distinct character with significant aesthetic, cultural and/or ecological value, and often with high biological diversity. Safeguarding the integrity of this traditional interaction is vital to the protection, maintenance and evolution of such an area.

VI Managed Resource Protected Area: Protected Areas Managed Mainly for the Sustainable Use of Natural Ecosystems. Areas containing predominantly unmodified natural systems managed to ensure long-term protection and maintenance of biological diversity, while providing at the same time a sustainable flow of natural products and services to meet community needs.

The IUCN guidelines on applying the categories emphasize that all categories are important and needed for conservation and sustainable development. All the countries are encouraged to develop a system of protected areas in a wide range of categories that meet their own natural and cultural heritage conservation objectives. Each category is defined by the "principal" objective of its management, which means two-thirds or more of a particular protected area is managed for that primary purpose. The rest of the area can accommodate activities that comply with fundamental management objective (ICEM, 2003). The purposes of this international classification are to reduce the confusion of terminology, to provide an agreed set of international management standards and to facilitate international comparison and accounting (Phillips and Harrison, 1997).

In Cambodia, the protected areas system includes 7 national parks (4 are coastal and marine protected areas), 10 wildlife sanctuaries, 3 protected landscapes, 3 multiple use areas (one of which is a coastal and marine area), 13 fish sanctuaries and 2 protected forests (ICEM, 2003). Map 1 below shows about the protected area system and the location of the protected areas in Cambodia. (More detail refers to table 1 in Annex). The national parks, wildlife sanctuaries, fish sanctuaries and protected forests conform to categories II and IV of IUCN's classification system, areas at least based on their stated objectives of management. The 3 protected landscapes conform to category V of the IUCN's classification system and the 3 multiple use areas to category VI. The designations of Biosphere Reserve, World Heritage Site and Ramsar site are not discrete management categories but titles given to areas of global importance under international agreements. They are generally overlain on one or other of the IUCN management categories (ICEM, 2003).

Map 1: Protected Area System in Cambodia

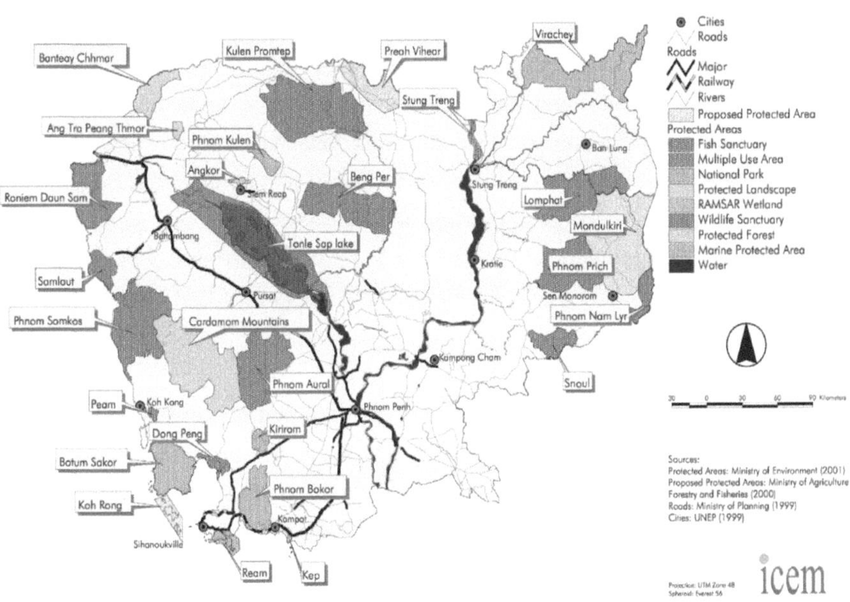

Source: ICEM, 2003

4. Barriers in protected area management and conservation

The main barriers in protected area management and conservation are: economic land concession, land encroachment and population pressure, illegal logging and hunting, climate change and capacity building.

4.1 Economic Land Concession

Agricultural sector was considered as one of the core pillar of country development thus the Royal Government of Cambodia start to introduce Economic Land Concession Scheme to contribute to economic growth of the country. Economic Land Concession encourages both domestic and international investors to gain access of state land for forestry and agro-industrial plantation. The objectives of this scheme are to increase employments in rural area, generate state revenue and develop agricultural sector (MAFF, 2007). State income from economic land concession can be generated via land rental, charges and taxes (NGO forum, 2005). The maximum period of an economic land concession is limited to 99 years (ICEM, 2003). According to Barney (2007), most of the land concessions target to plant rubber, oil palm and cashew nuts and areas are over 10 000 hectares according to land law. This economic land concession scheme has a large impact on local communities by reducing access to both forest resource and forestland (USAID, 2006). Furthermore, it also affects the protected areas and for example, Some forest areas in Botum Sakor national park were illegally cleared by Green Rich Company (Barney, 2007).The agro-industry plantations in and around Ream, Bokor and Kep National Parks are creating significant impact on those parks (ICEM, 2003).

4.2 Land encroachment and population pressure

In Cambodia, 95% of rural population relies on fuel wood for cooking (FAO, 1997; Top *et al*, 2004). Hence, the demand for fuel wood is increasing due to population growth and illegal fuel wood collection in the protected areas always occurs. Another issues regarding to protected area management is land encroachment (ICEM, 2003). In a densely populated area near by the protected area, local farmers try to clear more land for agricultural crops. Coastal areas are also threatened by encroachment of intensive shrimp ponds, repeated burning of Melaleuca areas, and to a lesser extent, fuelwood and timber cutting (UP-MSI *et al*, 2002).

4.3 Illegal logging and hunting

Illegal logging is one of the major challenges in protected areas management. Most of the protected areas in Cambodia suffer from both small and large scale illegal logging. Most of the protected areas are located close to the forest concessions and the illegal loggings inside those forest concessions are common (ICEM, 2003). Poor enforcement of protected areas law is the major issue which led to illegal logging activities (Le Billon, 1999). As a result, twenty percent of forest in protected areas was affected by illegal logging (Henderson, 1999, cited in ICEM, 2003). Another remarkable example was about illegal cutting down of *Cinnamomum parthenoxylon* (Mreah Prew Phnom tree) in Phnom Samkos Wildlife Sanctuary of Cardamom Mountains (Fauna & Flora International, 2009). Sassafras oil was distilled by boiling the root and trunk of the tree and can also be use to produce MDMA which commonly known as ecstasy. This distillation threatened the forest ecosystem in the protected area (Fauna & Flora International, 2009).The designation of protected area is due to the important of wildlife habitats (Ministry of Environment, 1999). As a result, illegal hunting is a major issue for protected area conservation. Furthermore, the high demand and lucrative market in Vietnam, Thailand, China, Hong Kong and Tawain are the driving force of illegal hunting of endangered species in Cambodia (Ministry of Environment, 1999; ADB, 2000).

4.4 Climate change

Climate change is the most challenge issues to ecosystem on the earth. Protected areas are inevitable to climate change impact (Shadie and Epps, 2008). Cambodia is one of the most venerable nations in Southeast Asia due to climate change impact on agriculture, coastal zones, water resources and health (UNDP, 2008). However, due to Goreau *et al* (2000) the coral reef in Cambodia has suffered from severe morality due to bleaching which caused by high positive sea surface temperature. It is likely to show that the marine ecosystem has already affected by the climate change. Another recent study by van Zonneveld *et al* (2009) illustrated that the changing climate was expected to exceed the tolerant of *pinus merkusii* (two needle leaves pine) which can lead to degradation and extinction. Noticeably, *pinus* merkusii is one of the most dominant and important species in Kirrirom National Park in Cambodia. Furthermore, this species is already listed in the vulnerable conservation status of IUCN (IUCN, 2008).

4.5 Capacity building and Financial Resource

Human resource skill and capacity improvement are the main components to formulate and implement the effective policies in managing protected areas (Ministry of Environment, 1999). The skill and knowledge of national park rangers are still limited to deal with complex management of park areas (Khim, 2000). However, the technical trainings are needed to build up the capacity of those rangers. These are the most challenges which need to be overcome to achieve sustainable management of the protected areas in Cambodia. Additionally, management of protected area require large budget to pay for the staff, to demarcate park boundaries so that communities are aware of the protected area, to carry out research on the plants and animals which inhabit the park, to undertake projects with communities which live in the buffer zones, to rehabilitate infrastructures in protected area, and to design brochures and displays for visitors to educate them about the park and its plants and animals and how they can help protect Cambodia's natural resources (Ministry of Environment, 1999; ICEM, 2003).

5. Opportunities in protected area management and conservation

The important opportunities in protected area management and conservation in Cambodia are: tourism, climate change mitigation and adaptation, scientific research, and transboundary cooperation.

5.1 Tourism

Many poor communities are living around the protected areas and depending on the resources from protected area (Morris and Vathana, 2003). Tourism is the second largest actor in Cambodian economy after garment industry (Chheang, 2008). In 2000, tourism revenues reached to US$20 million which accounted for 6.3 percent of total government revenues and in 2002, tourism contribution increased up to 10 percent of GDP (ICEM, 2003). Natural base tourism and ecotourism play an important role in generating job and income for the communities around the protected areas. As a result, promoting the both kinds of tourism have been made in Cambodia. For example, Malup Biatong, local NGO, decided to launch ecotourism project in Kirrirom National Park with the collaboration with the Ministry of Environment in order to promote local livelihood (Moeurn *et al*, 2008). The outcomes of this project have a significant contribution to reduce the poverty and enhance environmental conservation within the communities in the protected area. The protected landscape of Angkor Wat, one of the Eight Wonders of the World, is the most famous protected areas which attract many tourists around the world (Leksakundilok, 2004).

5.2 *Climate change mitigation and adaptation*

Promoting protected areas is also one of the important options for climate change adaptation (CBD, 2007).Cambodia ratified the United Nations Framework Convention on Climate Change (UNFCCC) in 1995 and acceded to its Kyoto Protocol in 2002 (Ministry of Environment, 2006). The flexible mechanisms under the Kyoto protocol like Clean Development Mechanism (CDM) would help can Cambodia to benefit from emission trading. Adding to Kyoto protocol, UNFCCC introduced Reduction Reducing Emissions from Deforestation and Forest Degradation (REDD) as the financial mechanism to reduce emission from deforestation and forest degradation in developing countries (Huettner *et al*, 2009; Miles and Kapos, 2009). Generally, most of the protected areas in Cambodia are covered by tropical rainforest. Hence, REDD is also seen as new tool for protected areas conservation measure.

5.3 *Scientific Research*

In Cambodia, existing knowledge about fauna and flora species is still limited because during the French colonization little record was made and the civil war also made the scientist not able to explore the Cambodian's forest (Ohler et al, 2002). More scientific research needs to be done especially in the protected areas. Recent scientific research in 2007 discovered a new species of rhacophorid frog, *Chiromantis samkosensis*, from Phnom Samkos of the Cardamom Mountains and this frog distinguished from other species of Asian *Chiromantis* by having green blood and turquoise bones amongst other unique characteristics (WWF, 2008). Furthermore, due to WWF (2008) new pocket size wolf snake *Lycodon cardamomensis* is also discovered the remote and least know part of Cardamom mountain range and in 2007, woolly bat *Kerivoula titania* was also identified in Seima Biodiversity Conservation Area. The researches in protected areas are vital for scientific communities to explore new species and better understand of ecosystem.

5.4 Transboundary cooperation

Cambodia, Laos and Vietnam have natural habitats at the international borders with high enough biodiversity values for transboundary conservation system (Parks for Peace Conference Proceedings, 1997). International collaboration within neighboring countries is an opportunity for strengthening biodiversity over the shared ecosystem (ICEM, 2003). To some extent, it is also help to facilitate the movement of migrant species in their natural range regardless the physical borders. For instance, The Virechey national park in Cambodia is contiguous with protected areas in Lao PDR and Vietnam so this park has a great potential for transboundary cooperation (ICEM, 2003).

6. Conclusion

Protected areas have played vital role in ecosystem conservation in Cambodia. The management of protected areas still is facing lot of challenges such as economic land concession, land encroachment and population pressure, illegal logging and hunting, climate change and capacity building. Furthermore, there are more barriers regarding to protected area management but only the most critical issues was taken into account. In spite of the barriers, there were also the opportunities like tourism, climate change mitigation and adaptation, scientific research, and transboundary cooperation which can help to achieve the effective management and conservation of the protected areas in Cambodia.

References

1. ADB (Asian Development Bank).2000: Environments in Transition Cambodia, Lao PDR, Thailand, Viet Nam. Manila, Philippine.
2. Barney, K. 2007. A Note on Forest Land Concessions, Social Conflicts, and Poverty in the Mekong Region. Proceedings: International Conference on Poverty Reduction and Forests, Bangkok, Thailand.
3. Billon, 1999: Environment: ETAP Reference Guide Book: Forestry issues of Cambodia.
4. Cambodian team, (2009, 19 August): Development of Protected Area Law & the Environmental Related Laws in Cambodia. PowerPoint Presentation.

International Workshop on "Experiences and Challenges in Biodiversity Policy Development in Greater Mekong Subregions Countries". Hanoi, Vietnam.

5. Carew-Reid, J.2003: Protected areas as engines for good governance and economic reform in the lower Mekong River regions. PARKS. Vol 13: No 3

6. CBD (Convention on Biological Diversity).2007: Biodiversity and climate change.

7. Chheang, V. 2008: The Political Economy of Tourism in Cambodia. Asia Pacific Journal of Tourism Research,13:3, 281 — 297

8. Dudley, N. (Editor).2008: Guidelines for Applying Protected Area Management Categories. International Union for Conservation of Nature (IUCN). Gland, Switzerland.

9. Fauna & Flora International (FFI). 2009: 'Ecstasy oil' distilleries raided in Cambodia's Cardamom Mountains. Media Release. Phnom Penh, Cambodia

10. Huettner, M., Leemans, R., Kok, K. and Ebeling, J. 2009: A comparison of baseline methodologies for 'Reducing Emissions from Deforestation and Degradation'. Carbon Balance and Management, 4:4 doi: 10.1186/1750-0680-4-4

11. ICEM. 2003: Cambodia National Report on Protected Areas and Development. Review of Protected Areas and Development in the Lower Mekong River Region, Indooroopilly, Queensland, Australia. 148 pp.

12. ITTO (International Tropical Timber Organization).2005: Status of Tropical Forest Management 2005: Cambodia.

13. International Union for Conservation of Nature (IUCN).2008: *2008* IUCN Red List of Threatened Species

14. FAO (Food and Agriculture Organization), 1997: The role of wood energy in Asia. Bangkok, Thailand.

15. FAO (Food and Agriculture Organization of United Nation). 2007: Cambodia

16. FAO (Food and Agriculture Organization of United Nation). 2009: Cambodia: Geographic description. (http://www.fao.org/forestry/country/18310/en/khm/)

17. Goreau, T. J. Hayes, R. L. and McClanahan, T. 2000: Conservation of Coral Reefs after the 1998 Global Bleaching Event. Conservation Biology, Vol. 14, No. 1

18. Khim, L. 2000: Biodiversity and Protected Areas Management Project: National Policy and Capacity Building Component Training Needs Assessment of Rangers of Virachey National Park, Kingdom of Cambodia.

19. Leksakundilok, A. 2004: Ecotourism and Community-based Ecotourism in the Mekong Region. Working Paper No. 10. Australian Mekong Resource Centre. University of Sydney.

20. MAFF (Ministry of Agriculture, Forestry and Fisheries). 2007: Economic Land Concession. Available at: http://www.elc.maff.gov.kh/index.html

21. Miles, L. and Kapos, V. 2009: Reducing Greenhouse Gas Emissions from Deforestation and Forest Degradation: Global Land-Use Implications. Science 320, 1454

22. Ministry of Rural Development. 2006: Information on Population of Cambodia. Available at: http://www.mrd.gov.kh/information/land_res.htm

23. Ministry of Environment.1999: Environment: Concepts and issues: A focus on Cambodia. UNDP/ETAP Reference guidebook. Phnom Penh, Cambodia

24. Ministry of Environment .2006: National Adaptation Programme of Action to Climate Change (NAPA). Phnom Penh, Cambodia

25. Moeurn,V. Khim, L and Sovanny, C. 2008: Good Practice in the Chambok Community-Based Ecotourism Project in Cambodia

26. Morris, J. and Vathana, K. 2003: Poverty reduction and protected areas in the Lower Mekong region. PARKS. Vol 13: No 3

27. NGO Forum. 2005: Fast-wood Plantations, Economic Concessions and Local Livelihoods in Cambodia: Field Investigations in Koh Kong, Kampong Speu, Pursat, Kampong Chhnang, Mondolkiri, Prey Veng and Svay Rieng Provinces

28. Ohler, A.,Swan, S.R.,Daltry, J.C., 2002: Recent Survey of the Amphibian Fauna of the Cardamon Mountains, Southwest Cambodia with Descriptions of Three New Species. The Raffles Bulletin of Zoology 50(2): 465-481. Singapore.

29. Parks for Peace Conference Proceedings, 1997: Park for Peace. International conference on tranboundary protected areas as a vehicle for international co-operation.

30. Phillips, A and Harrison, J. 1997: The framework for International Standards in Establishing National Parks and Other Protected Areas. The George Wright Forum, Vol. 14, No 2

31. Shadie, P. and Epps, M. (Eds) (2008). Securing Protected Areas in the Face of Global Change: Key lessons learned from case studies and field learning sites in protected areas. IUCN Asia Regional Office, Bangkok, Thailand. 49pp

32. Top, N., Mizoue, N., Kai, S., Nakao, T. 2004: Variation in wood fuel consumption patterns in response to forest availability in Kampong Thom Province, Cambodia. Biomass and Bioenergy 27: 57 – 68

33. USAID (United States Agency for International Development). 2006: Forest Conflict in Asia: Undermining Development, Security, and Human Rights.

34. UNDP (United Nation Development Programme).2008: UNDP Annual Report 2008. Phnom Penh, Cambodia.

35. UP-MSI, ABC, ARCBC, DENR, ASEAN, 2002. Marine Protected Areas in Southeast Asia. ASEAN Regional Centre for Biodiversity Conservation, Department of Environment and Natural Resources, Los Baños, Philippines.

36. van Zonneveld, M. Koskela, J. Vinceti, B. and Jarvis, A. 2009: Impact of climate change on the distribution of tropical pines in Southeast Asia. Unasylva No. 231/232 Vol. 60, Rome, FAO.

37. Wager, J.1995: Developing a strategy for the Angkor World Heritage Site. Tourism Management, Vol. 16, No. 7, pp. 515-523

38. Worldbank.2007: World Development Report 2008: Agriculture for Development. Washington DC, United State of America

39. WWF (World Wildlife Fund), 2008: First Contact in the Greater Mekong. Hanoi, Vietnam

40. Yusuf, A.A. and Francisco, H.A. 2009: Climate Change Vulnerability Mapping for Southeast Asia. Economy and Environment Program for Southeast Asia (EEPSEA). Singapore.

Annex

Table 1: Protected Area in Cambodia

(Forestry Administration, www.forestry.gov.kh/Statistic/Forestcover.htm)

Protected Areas	Province/Municipality	Designated by	Area (ha)
National Parks			
Kirirom	Kampong Speu, Koh Kong	Royal Decree 1993	35,000
Phnom Bokor	Kampot	Royal Decree 1993	140,000
Kep	Kampot	Royal Decree 1993	5,000
Ream	Sihanouk Ville	Royal Decree 1993	21,000
Botum Sakor	Koh Kong	Royal Decree 1993	171,250
Phnom Kulen	Siem Reap	Royal Decree 1993	37,500
Virachey	Stung Treng, Ratanak Kiri	Royal Decree 1993	332,500
Wildlife Sanctuaries			
Phnom Aural	Koh Kong, Pursat, Kg Chhnang	Royal Decree 1993	253,750
Peam Krasop	Koh Kong	Royal Decree 1993	23,750
Phnom Samkos	Koh Kong	Royal Decree 1993	333,750
Roniem Daun Sam	Battambang	Royal Decree 1993	40,021
Kulem Promtep	Siem Reap, Preah Vihear	Royal Decree 1993	402,500
Boeng Per	Kampong Thom	Royal Decree 1993	242,500
Lomphat	Ratanak Kiri, Mondul Kiri	Royal Decree 1993	250,000
Phnom Prich	Mondul Kiri, Kratie	Royal Decree 1993	222,500
Phnom Namlear	Mondul Kiri	Royal Decree 1993	47,500
Snoul	Kratie	Royal Decree 1993	75,000
Protected Landscapes			
Angkor	Siem Reap	Royal Decree 1993	10,800
Banteay Chmar	Banteay Meanchey	Royal Decree 1993	81,200
Preah Vihear	Preah Vihear	Royal Decree 1993	5,000
Multiple Use Areas			
Dong Peng	Koh Kong	Royal Decree 1993	27,700
Samlaut	Battambang	Royal Decree 1993	60,000
Tonle Sap	Kampong Chhnang, Pursat, Siem Reap, Battambang, Kampong Thom	Royal Decree 1993	316,250

Protected Areas Total   **3,134,471**

Map 2: Climate change vulnerable of Southeast Asia (Yusuf and Francisco, 2009)

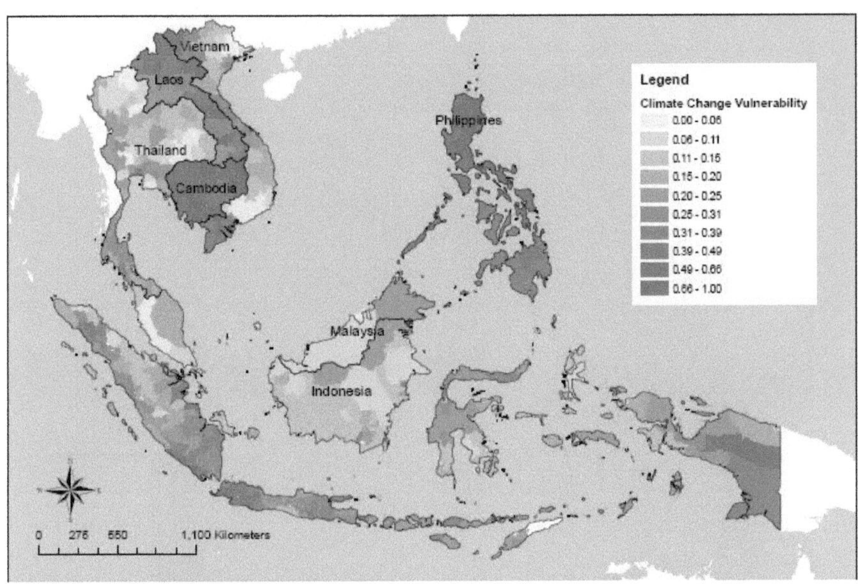

Note: For the legend, the scale used is 0-1 indicating the lowest vulnerability level (0) to the highest vulnerability level (1).

Map 3: Forest Cover of Cambodia (Forestry Administration, 2002)

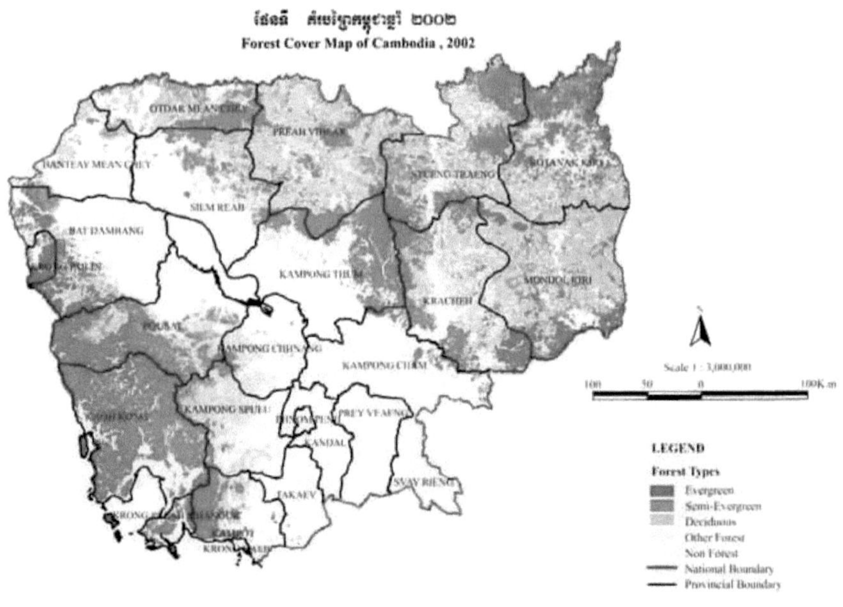